This study material has been produced by using the resource available in © State of New South Wales (Department of Education), 2023 as a guide. The author does not have any connection with © State of New South Wales (Department of Education) and this work has not been endorsed by © State of New South Wales (Department of Education).

1. If a watch reads 2220. What is the time one quarter of an hour before, in 12-hour format?

    A 10:05 am

    B 10:25 am

    C 9:55 pm

    D 10:35 pm

    E 10:05 pm

2. If a watch reads 0010. What is the time three quarter of an hour before, in 12-hour format?

    A 10:05 am

    B 10:05 am

    C 9:35 pm

    D 10:35 pm

    E 11:25 pm

3. If a watch reads 1310. What is the time One and half of an hour before, in 12-hour format?

    A 10:05 am

    B 11:40 am

    C 9:35 am

    D 10:35 am

    E 11:25 am

4. If a watch reads 1310 what is the time Two and half of an hour after, in 12-hour format?

    A 3:40 pm

    B 11:40 am

    C 12:40 pm

    D 10:40 am

    E 11:25 am

5. A watch reads 1310. If the watch runs three times faster than the normal watch. What is the time quarter of an hour after, in 12-hour format?

   A 1:55 am

   B 1:25 pm

   C 1:55 pm

   D 1:25 am

   E 12:25 pm

6. Mat divides a white wall into equal sections. He paints some sections blue, as shown. Mat then paints 25 % of the remining wall yellow. Then he paints ⅔ of the remaining wall with red. Finally, he paints the rest of the wall black. How many sections does he paint black?

   A 2
   B 8
   C 6
   D 4
   E 10

7. Mat divides a white wall into equal sections. He paints some sections blue, as shown. Mat then paints 25 % of the remining wall yellow. Then he paints 5/6 of the remaining wall with red. Finally, he paints the rest of the wall black. How many sections does he paint black?

   A 2
   B 8
   C 6
   D 4
   E 10

8. Mat divides a white wall into equal sections. He paints some sections blue, as shown. Mat then paints 75 % of the remining wall yellow. Then he paints ¾ of the remaining wall with red. Finally, he paints the rest of the wall black. How many sections does he paint black?

A 2
B 1
C 3
D 4
E 5

9. Mat divides a white wall into equal sections. He paints some sections blue, as shown. Mat then paints 50 % of the remining wall yellow. Then he paints ¾ of the remaining wall with red. Finally, he paints the rest of the wall black. How many sections does he paint black?

A 2
B 1
C 3
D 4
E 5

10. Mat divides a white wall into equal sections. He paints some sections blue, as shown. Mat then paints 12.5 % of the remining wall yellow. Then he paints 6/7 of the remaining wall with red. Finally, he paints the rest of the wall black. How many sections does he paint black?

A 2
B 1
C 3
D 4
E 5

11. Nina multiplies the smallest three-digit whole number by the largest one-digit whole number. What is the answer?

A 9000

B 900

C 90

D 90000

E 9900

12. Lassi divides the largest three-digit whole number by the smallest two-digit whole number. What is the answer?

   A 99.9

   B 9.99

   C 999

   D 0.999

   E 9.09

13. Cupid multiplies the smallest three-digit whole ODD number by the largest one-digit whole EVEN number. What is the answer?

   A 8080

   B 800

   C 808

   D 80000

   E 8800

14. Pilki divides the largest three-digit whole EVEN number by the smallest two-digit whole EVEN number. What is the answer?

   A 99.8

   B 9.98

   C 998

   D 0.998

   E 9.08

15. Aeriel ADDS the largest three-digit whole number to the smallest two-digit whole number. What is the answer?

   A 1019

   B 998

   C 1009

D 999

E 1090

16. Anh SUBSTRACTS the largest two-digit whole number by the smallest two-digit whole number. What is the answer?

   A 89

   B 88

   C 109

   D 990

   E 0

17. Simon needs 450 grams of cooking sauce. The scale is faulty, and each gram is represented as 3 grams. She places a glass jar of cooking sauce on a scale, and it shows 810 grams. If the real weight of the empty glass jar is 120 grams in a correct scale how much more cooking sauce does Rana need?

   A 300 g

   B 350 g

   C 200 g

   D 400 g

   C 100 g

18. Simon needs 390 grams of cooking sauce. The scale is faulty and each gram is represented as 0.5 grams. She places a glass jar of cooking sauce on a scale, and it shows 250 grams. If the real weight of the empty glass jar is 120 grams in a correct scale how much more cooking sauce does Rana need?

   A 10 g

   B 20 g

   C 30 g

   D 40 g

   C 100 g

19. Simon needs 390 grams of cooking sauce. The scale is faulty and each gram is represented as 0.75 grams. She places a glass jar of cooking sauce on a

scale, and it shows 225 grams. If the real weight of the empty glass jar is 120 grams in a correct scale how much more cooking sauce does Rana need?

A 210 g

B 200 g

C 230 g

D 240 g

C 100 g

20. Simon needs 390 grams of cooking sauce. The scale is faulty, and each gram is represented as 1.25 grams (g). She places a glass jar of cooking sauce on a scale, and it shows 625 grams. If the real weight of the empty glass jar is 120 grams in a correct scale how much more cooking sauce does Rana need?

A 10 g

B 200 g

C 30 g

D 40 g

C 100 g

21. Simon needs 390 grams of cooking sauce. The scale is faulty, and each gram is represented as 1.75 grams. She places a glass jar of cooking sauce on a scale, and it shows 1050 grams. If the real weight of the empty glass jar is 120 grams in a correct scale how much sauce does Rana needs to remove from the bottle?

A 90 g

B 110 g

C 130 g

D 80 g

C 100 g

22. X and X having 2 lines of symmetry as shown below.

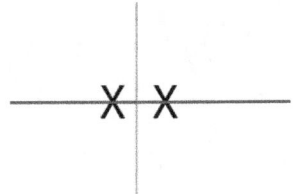

What is the largest two-digit number having 2 lines of symmetry

A. 90

B. 99

C. 96

D. 88

E. 86

23. X and X having 2 lines of symmetry as shown below.

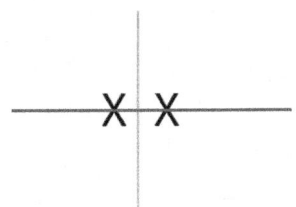

Which is the largest number from the below that is having 1 line of symmetry

A. 90

B. 99

C. 96

D. 88

E. 86

24. X and X having 2 lines of symmetry as shown below.

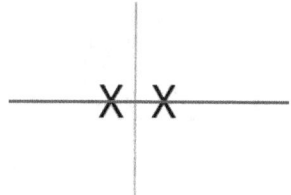

Which is the largest number from the below that is having NO line of symmetry

A. 98

B. 99

C. 96

D. 88

E. 86

25. Which shape of the below is incorrect about the line of symmetry?

A. Pentagon has 5 lines of symmetry.

B. Rhombus has 2 lines of symmetry

C. Kite has 1 lines of symmetry

D. Trapezium has 0 lines of symmetry

E. Parallelogram has 0 lines of symmetry

26. How many lines of symmetry does the rectangle have?

A. 1

B. 2

C. 4

D. 8

E. 0

27. Katia and Sylvia each think of a different whole number that is greater than zero and less than 100. Katia's number is a multiple of 9. Sylvia's number is an even number which is a multiple of 7. What is the difference between the largest possible value of Katia's number and the smallest possible value of Sylvia's number?

A 78

B 83

C 85

D 92

E 93

28. Katia and Sylvia each think of a different whole number that is greater than 3 and less than 30. Katia's number is a multiple of 6. Sylvia's number is an even number which is a multiple of 3. What is the difference between the largest possible value of Katia's number and the smallest possible value of Sylvia's number?

A 18

B 24

C 27

D 21

E 15

29. Katia and Sylvia each think of a different whole number that is greater than 3 and less than 60. Katia's number is a multiple of 7. Sylvia's number is an even number which is a multiple of 5. What is the difference between the largest possible value of Katia's number and the largest possible value of Sylvia's number?

A 6

B 1

C 106

D 10

E 5

30. Katia and Sylvia each think of a different whole number that is greater than 3 and less than 100. Katia's number is a multiple of 3 and it is a half of Sylvia's number. Sylvia's number is an even number which is a multiple of 5. What is the difference between the value of Katia's number and the largest possible value of Sylvia's number?

A 45

B 90

C 30

D 15

E 5

X

31. There are 2 different rectangles with same perimeter. One rectangle has sides 5cm and 10 cm. The second having one side 7cm and the other side is not given. What is the length of the other side of the 2nd rectangle?

A 8

B 10

C 6

D 4

E 12

32. There is a rectangle and an equilateral triangle with same perimeter. The rectangle has sides 5cm and 10 cm. What is the length of each side of the triangle?

A 8

B 10

C 6

D 4

E 12

33. There is a regular hexagon (with all sides with the same length) and an regular nonagon (with all sides with the same length) with same perimeter. The hexagon is having each side length of 15 cm. What is the length of each side of the nonagon?

A 8

B 10

C 6

D 20

E 12

34. There is a square with each side 5 cm and a second object having each side length of 2.5cm. If both objects having the same perimeter what shape would be the second object.

A hexagon

B pentagon

C rectangle

D Octagon

E Dodecagon

35. There is a regular nonagon with each side length of 10 cm. How many squares with 2.25cm each side length will be required to have the same perimeter pf the nonagon?

A 36

B 18

C 9

D 10

E 40

36. There are 2 different rectangles (rectangle A and rectangle B) with the same area. Rectangle A is having sides 5cm and 10 cm. The second rectangle having one side 25 cm, and the other side is not given. What is the length of the other side of the rectangle B?

A 2

B 1

C 20

D 4

E 5

37. There are 2 different rectangles (rectangle A and rectangle B) with the same area. Rectangle A is having sides 7cm and 12 cm. The second rectangle having one side 21 cm, and the other side is not given. What is the length of the other side of the rectangle B?

A 2

B 1

C 40

D 4

E 5

38. There is a square with each side is 10cm. what would be the side of 25 small squares to have the same area?

A 2cm

B 6cm

C 1 cm

D 3 cm

E 5 cm

39. Area of the 4 identical squares with each side of 10 cm is same as the area of 5 identical rectangles. If the length of one side is 20% of the other side what would be the length of each side?

A 20cm and 4 cm

B 1cm and 4 cm

C 4cm and 8 cm

D 3cm and 6 cm

E 5cm and 10 cm

40. Area of the 6 identical squares with each side of 10 cm is same as the area of 2 identical rectangles. If the length of one side is ⅓ of the other side what would be the length of each side?

A 2cm and 6 cm

B 10cm and 30 cm

C 3cm and 9 cm

D 4cm and 12 cm

E 5cm and 15 cm

41. Sakiyo has a glass containing 200 millilitres of cordial, and a bottle containing 1 litre of water. Sakiyo drinks 20 millilitres of the cordial. She finds the drink too strong, so she pours water from the bottle into the glass until the water reaches the 500 millilitres mark. How much water is left in the bottle?

A 730ml

B 700ml

C 680ml

D 600ml

E 540ml

42. Sakiyo has a glass containing 200 millilitres of orange cordial, a glass containing 100 millilitres of lemon cordial and a bottle containing 1 litre of water. Sakiyo drinks 20 millilitres of each cordial. She decided to pour the remaining lemon cordial to the orange cordial glass. She thought it was too strong and decided to empty ½ of the mixed cordial and so she pours water from the bottle into the glass until the water reaches the 500 millilitres mark. How much water is left in the bottle?

A 730ml

B 700ml

C 630ml

D 600ml

E 540ml

43. Sakiyo has a glass containing 150 millilitres of orange cordial, a glass containing 100 millilitres of lemon cordial and a bottle containing 1 litre of water. Sakiyo drinks ½ of each cordial. She decided to pour the remaining lemon cordial to the orange cordial glass. She thought it was too strong and decided to empty ½ of the mixed cordial and so she pours water from the bottle into the glass until the water reaches the 700 millilitres mark. How much water is left in the bottle?

A 362.5ml

B 372.5ml

C 352.5ml

D 452.5ml

E 552.5ml

44. Sakiyo has a glass containing 250 millilitres of orange cordial, a glass containing 150 millilitres of lemon cordial and a bottle containing 1 litre of water. Sakiyo drinks 1/5 of each cordial. She decided to pour the remaining lemon cordial to the orange cordial glass. She thought it was too strong and decided to empty ½ of the mixed cordial and so she pours water from the bottle into the glass until the water reaches the 700 millilitres mark. How much water is left in the bottle?

A 460ml

B 350ml

C 360ml

D 450ml

E 550ml

45. Sakiyo has a glass containing 250 millilitres of orange cordial, a glass containing 150 millilitres of lemon cordial, a glass containing 100 millilitres of mango cordial and a bottle containing 1 litre of water. Sakiyo drinks 1/5 of each cordial. She decided to mix all 3 glasses. She thought it was too strong and decided to empty ½ of the mixed cordial and so she pours water from the bottle into the glass until the water reaches the 700 millilitres mark. How much water is left in the bottle?

A 460ml

B 500ml

C 360ml

D 450ml

E 550ml

46. □ always represents the same number. ▲ and @ represents different numbers. If 7 × □ = 63, @ x □ =36 and ▲ + □ -@= 36

what is ▲?

A 31

B 32

C 34

D 28

E 18

47. □ always represents the same number. ▲ and @ represent different numbers.
If 7 × □ = 63, @ x □ =36 and ▲ - □ -@= 36   what is ▲?

A 23

B 32

C 34

D 28

E 49

48. □ always represents the same number. ▲ and @ represents a different number.
If 6 × □ = 48, @ x □ =56 and ▲ + □ +@= 36

what is ▲?

A 21

B 32

C 34

D 28

E 18

49. □ always represents the same number. ▲ @ represent a different number. If 2 × □ × @= 112, @ x □ =▲ and ▲ + □+@= 71 ,□-@= 1

what is @?

A 7

B 6

C 5

D 4

E 8

50. □ always represents the same number. and @ represent a different number. If 4 x □ × @= 112, @ x □ =▲ and ▲ + □+@= 39 ,□-@= 3

what is @?

A 2

B 1

C 4

D 11

E 23

51. Toto is making a pattern out of white and blue tiles. She hasn't finished yet. She wants the finished pattern to have one vertical line of symmetry. What is the smallest number of tiles she needs to add onto the right of the pattern?

A 4
B 8
C 6
D 10
E 2

52. Toto is making a pattern out of white and blue tiles. She hasn't finished yet. She wants the finished pattern to have one vertical line of symmetry. What is the smallest number of tiles she needs to add onto the right of the pattern?

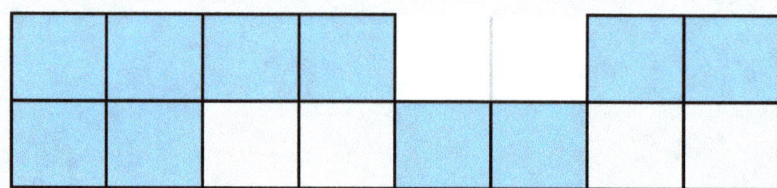

A 4
B 8
C 6
D 10
E 2

53. Toto is making a pattern out of white and yellow tiles. She hasn't finished yet. She wants the finished pattern to have one horizontal line of symmetry. What is the smallest number of yellow tiles she needs to add onto the right of the pattern?

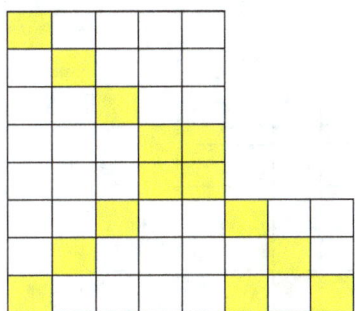

A 4
B 3
C 2
D 5
E 6

54. Mela is making a pattern out of white and yellow tiles. She hasn't finished yet. She wants the finished pattern to have two diagonal line of symmetry. What is the smallest number of yellow tiles she needs to add onto the pattern?

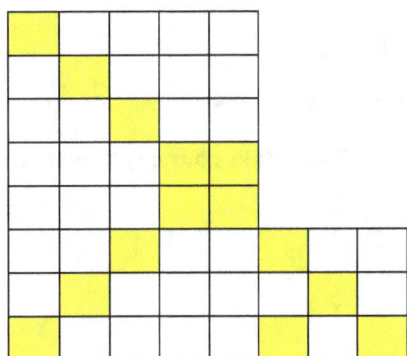

A 4
B 3
C 10
D 5
E 6

55. Toto is making a pattern out of white and yellow tiles. She hasn't finished yet. She wants the finished pattern to have horizontal, vertical and two diagonal line of symmetry. What is the smallest number of yellow tiles she needs to add onto the pattern?

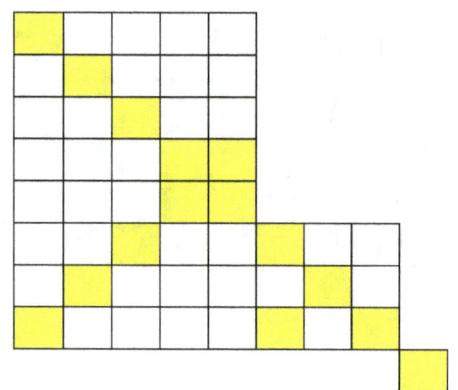

A 4
B 10
C 6
D 13
E 8

56. Karia has blue, green, red and yellow balls in a bag. There are three times as many red balls as blue balls. If she takes one ball out of the bag without looking

- the probability that it is blue is 0.1
- the probability that it is green is 0.4

What is the probability that Karia takes out a yellow ball?

A 0.1
B 0.3
C 0.5
D 0.7
E 0.2

57. Katy has blue, green, red and yellow balls in a bag. There are ½ as many red balls as blue balls. If she takes one ball out of the bag without looking:

• the probability that it is blue is 0.4

• the probability that it is green is 0.3

What is the probability that Katy takes out a yellow ball?

A 0.1

B 0.3

C 0.5

D 0.7

E 0.2

58. Bingo has blue, green, red and yellow balls in a bag. There are $\frac{1}{4}$ as many red balls as blue balls. If she takes one ball out of the bag without looking:

• the probability that it is blue is 0.4

• the probability that it is green is 0.3

What is the probability that Katy takes out a yellow ball?

A 0.1

B 0.3

C 0.5

D 0.7

E 0.2

59. Bluey has blue, green, red and yellow balls in a bag. There are $\frac{3}{4}$ as many red balls as blue balls. If she takes one ball out of the bag without looking:
• the probability that it is blue is 0.4
• the probability that it is green is 0.2

If the number of yellow balls in the bag is 10 what is the total number of the balls in the bag?

A 10

B 30

C 50

D 20

E 100

60. Emma has blue, green, red and yellow balls in a bag. There are half as many red balls as blue balls. If she takes one ball out of the bag without looking:
Number of yellow balls in the bag is 12.

• the probability that it is blue is 0.2

• the probability that it is green is 0.3

Then she added number of red balls, so the number of red balls equals the blue balls.

What is the number of balls in the bag once additional red balls were added?

A 33
B 30
C 27
D 36
E 39

61. The graph shows the average monthly rainfall in Sydney for one year

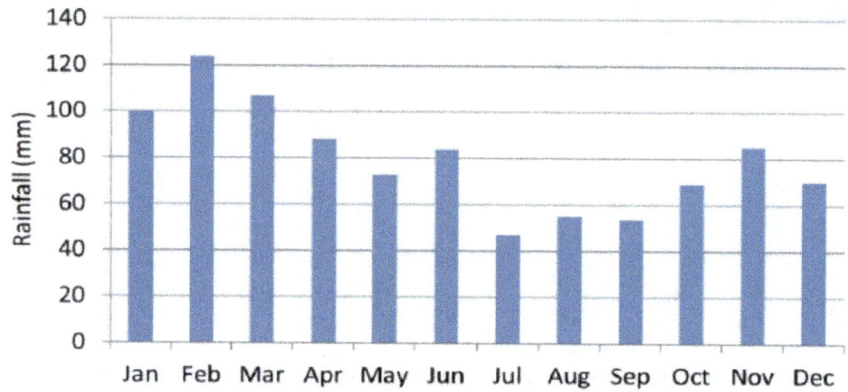

Average monthly rainfall in the Greater Sydney region.

Here are three statements about the graph.

1 February had more than twice as much rainfall as August.

2 January had about 10 cm of rain.

3 The highest rainfall was in February.

4 Lowest rainfall was in July.

of these statements are correct?

A statement 1 only

B statement 2 only

C statements 1 and 2 only

D statements 2 and 3 only

E All statements correct.

62. The graph shows the average monthly rainfall in Sydney for one year

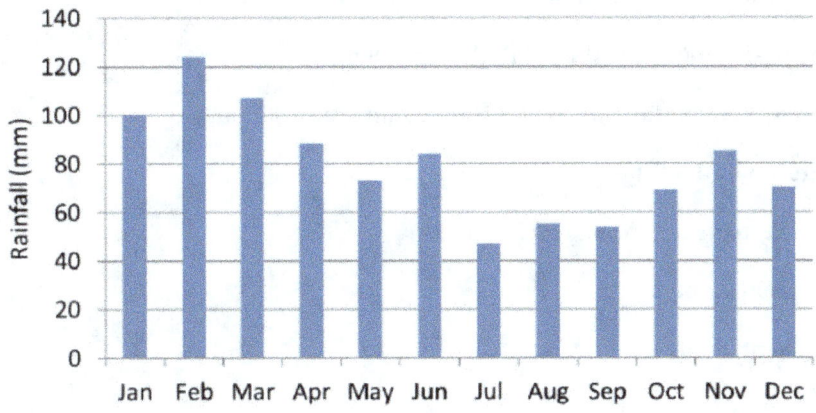

Average monthly rainfall in the Greater Sydney region.

Here are three statements about the graph.

1 February had more than twice as much rainfall as August.

2 October and December had similar rainfall.

3 First half of the year had more rain than the second half of the year.

4 June and November had similar rainfall.

of these statements are correct?

A statement 1 only

B statement 2 only

C statements 1 and 2 only

D statements 2 and 3 only

E All statements correct.

63. The graph shows the average monthly rainfall in Sydney for one year

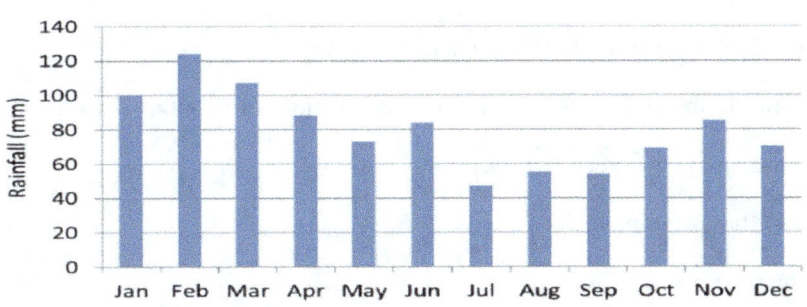

Average monthly rainfall in the Greater Sydney region.

Here are three statements about the graph.

1 All the months had rainfall over 40mm.

2 None of the months had a rainfall over 140 mm

3 The average rainfall has to be within 50mm and 100mm

4 Wettest month of the year is February..

of these statements are correct?

A statement 1 only

B statement 2 only

C statements 1 and 2 only

D statements 2 and 3 only

E All statements correct.

64. The graph shows the average monthly rainfall in Sydney for one year

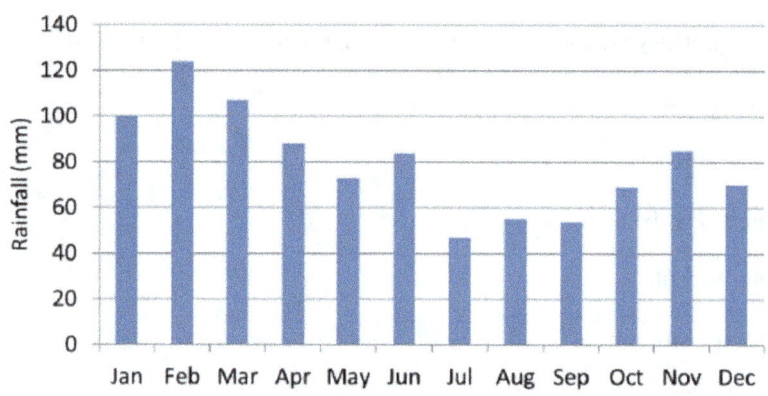

Average monthly rainfall in the Greater Sydney region.

Here are three statements about the graph.

1 April had close to twice as much of rainfall as July.

2 Second highest rainfall for the year is in March

3 The rainfall in March is higher than January and lower than February.

4 Driest month of the year is July.

Which of these statements are correct?

A statement 1 only

B statement 2 only

C statements 1 and 2 only

D statements 2 and 3 only

E All statements correct.

65. The graph shows the average monthly rainfall in Sydney for one year

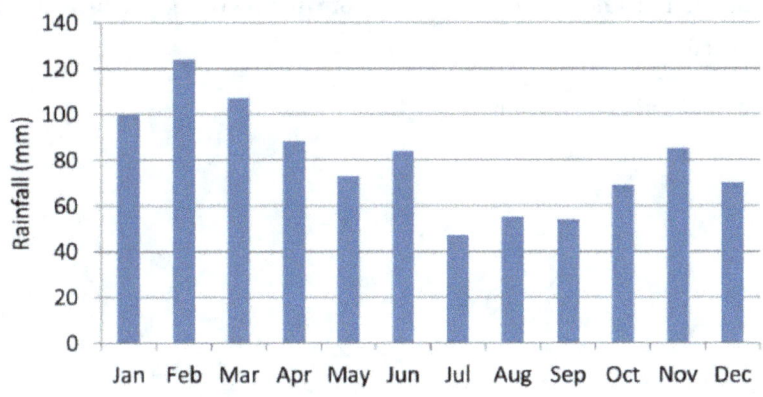

Average monthly rainfall in the Greater Sydney region.

Here are three statements about the graph.

1 Majority of the months had the rainfall over 60mm.

2 Only one month had the rainfall above 120mm

3 None of the months had rainfall below 20mm

4 3 months had rainfall below 60mm.

of these statements are correct?

A statement 1 only

B statement 2 only

C statements 1 and 2 only

D statements 2 and 3 only

E All statements correct.

66. Max had $20. He gave some of this money to Kona. Pat then gave half of what he had left to Jake. Jake gave half of what he was given to Jenny. Jenny received $1.60.

How much did Max give to Kona?

A $13.20

B $13.60

C $14.80

D $16.40

E $16.80

67. Pat had $30. He gave some of this money to Kona. Pat then gave quarter of what he had left to Jake. Jake gave quarter of what he was given to Jenny. Jenny received $1.60.

How much did Pat give to Kona?

A $13.20

B $13.60

C $4.80

D $4.40

E $16.80

68. Pat had $30. He gave some of this money to Kona. Pat then gave 12.5% of what he had left to Jake. Jake gave 12.5% of what he was given to Jenny. Jenny received $0.20.

How much did Pat give to Kona?

A $13.20

B $15.60

C $17.20

D $17.40

E $16.80

69. Spike had some money. He gave $\frac{1}{2}$ of this money to Kemon. Spike then gave 20% of what he had left to Jake. Jake gave 12.5% of what he was given to Jenny. Jenny received $0.20.

How much did Spike give to Kemon?

A $4

B $8

C $1

D $2

E $16

70. Spikey had some money. He gave 40% of this money to Wolfy. Spikey then gave 20% of what he had left to Mania. Mania gave 12.5% of what he was given to Jake. Jake received $0.30.

How much did Spikey have at the start?

A $40

B $80

C $10

D $20

E $160

71. What is the size of angle A?

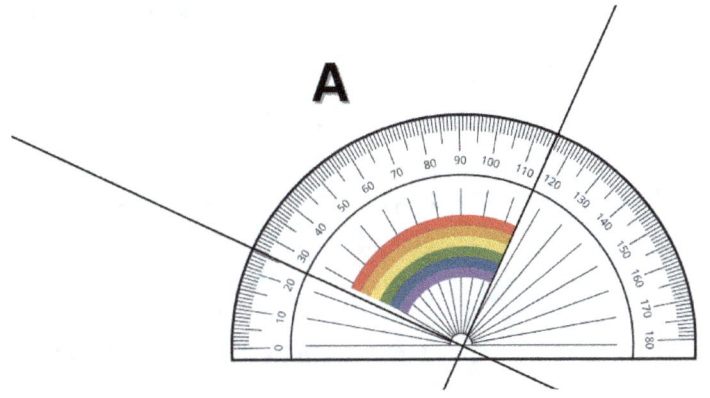

A 50°
B 80°
C 90°
D 100°
E 130°

72. If the angle A is divided by 6 and multiplied by 4 what is the angle?

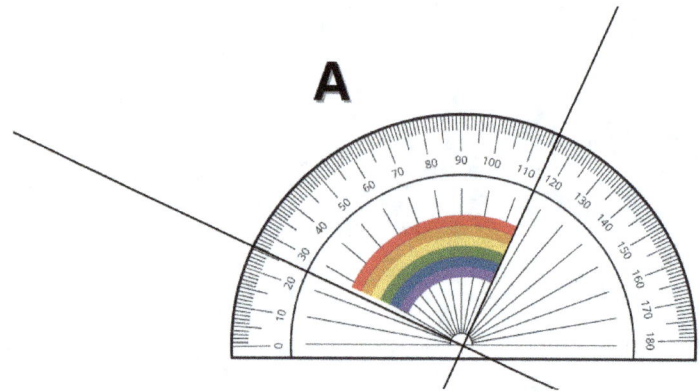

A 50°

B 80°

C 90°

D 60°

E 40°

73. If the angle A is divided by 2 and multiplied by 8 what is the angle?

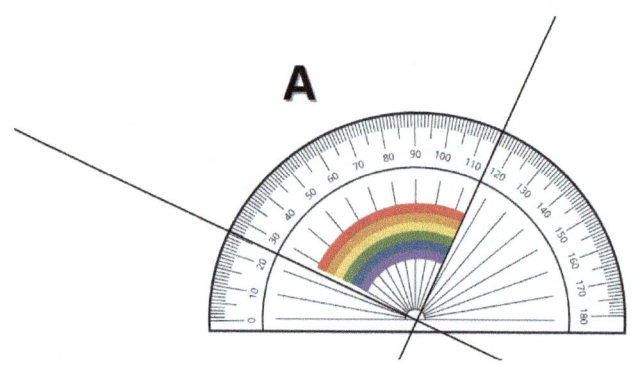

A 180°

B 160°

C 90°

D 360°

E 40°

74. If the angle A is divided by 2 what is the angle?

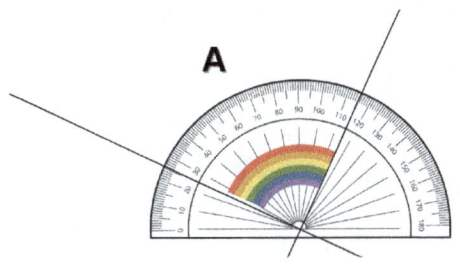

A Acute

B Right

C Obtuse

D Straight

E Reflex

75. If the angle A is divided by 2 and multiplied by 3 what is the angle?

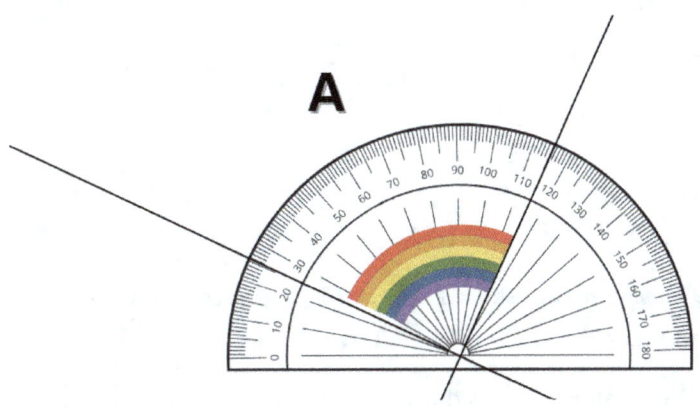

A Acute

B Right

C Obtuse

D Straight

E Reflex

76. When 468 chocolates are shared between 19 people, each person gets 24 chocolates and there are 12 chocolates left over.

Which number sentence shows this information?

A 24 + 19 × 12 = 468

B 24 × 12 – 19 = 468

C 24 × 12 + 19 = 468

D 24 × 19 – 12 = 468

E 24 × 19 + 12 = 468

77. When 112 chocolates are shared between 12 people, each person gets 9 chocolates and there are 4 chocolates left over.

Which number sentence shows this information?

A 4 + 9 × 12 = 112

B 4 × 12 – 9 = 112

C 4 × 12 + 9 = 112

D 4 × 9 – 12 = 112

E 4 × 9 + 12 = 112

78. When 99 chocolates are shared between 7 people, each person gets 14 chocolates and there are 1 chocolate left over.

Which number sentence shows this information?

A 1 + 7 × 14 = 99

B 1 × 14 – 7 = 99

C 1 × 14 + 7 = 99

D 1 × 7 – 14 = 99

E 1 × 7 + 14 = 99

79. When 111 chocolates are shared between 5 people, each person gets 22 chocolates and there are 1 chocolate left over.

Which number sentence shows this information?

A 1 + 5 × 22 = 111

B 1 × 22 – 5 = 111

C 1 × 22 + 5 = 111

D 1 × 5 – 22 = 111

E 1 × 5 + 22 = 111

80. When 888 chocolates are shared between 78 people, each person gets 11 chocolates and there are 30 chocolates left over.

Which number sentence shows this information?

A 30 + 11 ×78 = 888

B 30 × 11 – 5 = 888

C 30 × 11 + 5 = 888

D 30 × 11 – 78 = 888

E 30 × 11 + 78 = 888

81.

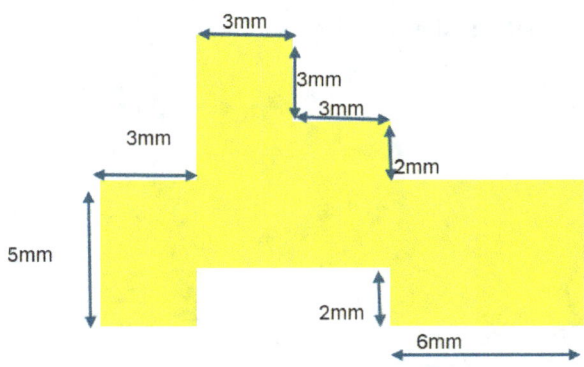

DIAGRAM IS NOT TO SCALE

A large rectangle has sides 15 mm and 10 mm. Some rectangles and a square are removed from this large rectangle. This leaves the shaded shape. What is the perimeter of the shaded shape?

A 30 mm

B 36 mm

C 39 mm

D 40 mm

E 54 mm

82.

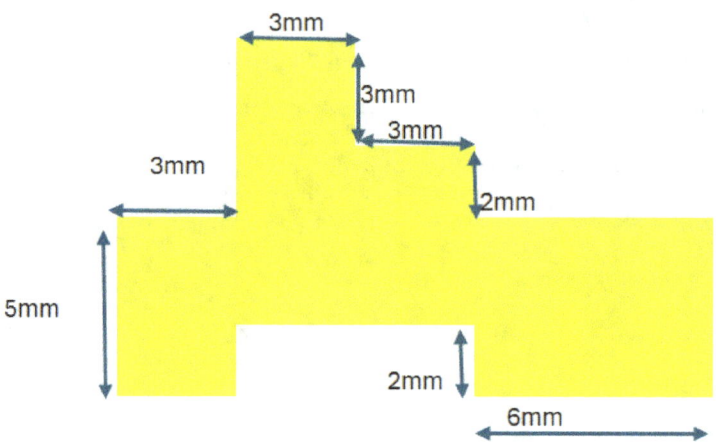

DIAGRAM IS NOT TO SCALE

A large rectangle has sides 15 mm and 10 mm. Some rectangles and a square are removed from this large rectangle. This leaves the shaded shape. What is the perimeter of the removed rectangles and a square?

A 72 mm

B 66 mm

C 39 mm

D 40 mm

E 54 mm

83.

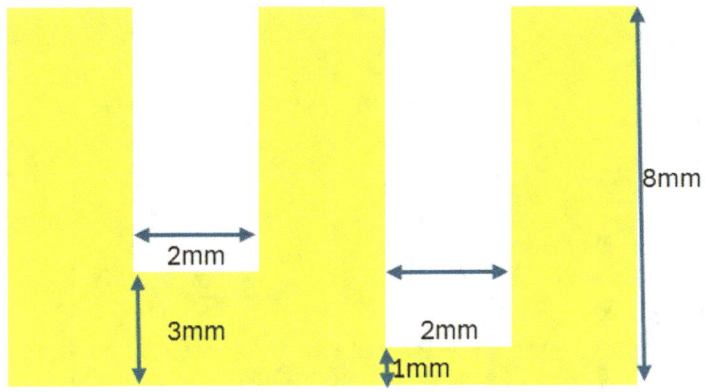

DIAGRAM IS NOT TO SCALE

A large rectangle has sides 8 mm and 10 mm. 2 rectangles are removed from this large rectangle. This leaves the shaded shape. What is the perimeter of the removed rectangles?

A 68 mm

B 32 mm

C 39 mm

D 40 mm

E 54 mm

84.

DIAGRAM IS NOT TO SCALE

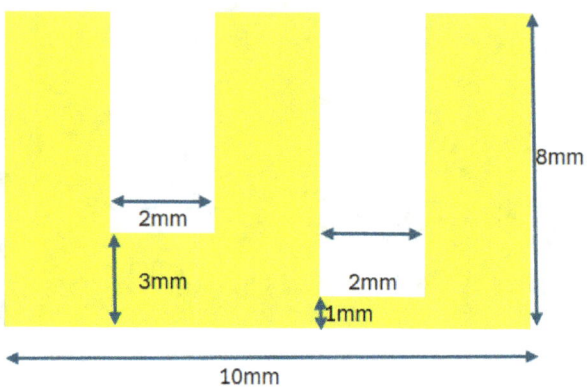

A large rectangle has sides 8 mm and 10 mm. 2 rectangles are removed from this large rectangle. This leaves the shaded shape. What is the perimeter of the shape?

A 60 mm

B 64 mm

C 69 mm

D 50 mm

E 54 mm

85.

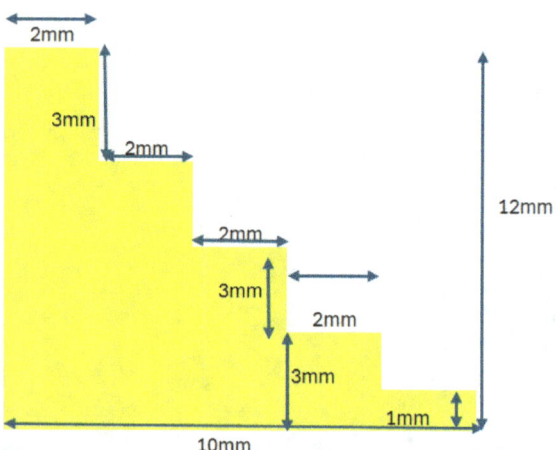

DIAGRAM IS NOT TO SCALE

A large rectangle has sides 8 mm and 10 mm. rectangles are removed from this large rectangle. This leaves the shaded shape. What is the perimeter of the shape?

A 60 mm

B 44 mm

C 69 mm

D 50 mm

E 54 mm

86.

Andy, Bella and Carlo shared a chocolate bar.

Andy ate 3/5 of the bar, Bella ate 1/5 of the bar and Carlo ate the rest. Which of the following statements are correct?

1 Carlo ate less than half of the chocolate bar.

2 Andy ate more than one quarter of the chocolate bar.

3 Andy and Bella ate less than three quarters of the chocolate bar altogether.

A statement 1 only

B statement 2 only

C statement 3 only

D statements 1 and 2 only

E statements 1 and 3 only

87. Andy, Bella and Carlo shared a chocolate bar.

Andy ate 1/2 of the bar, Bella ate 1/5 of the bar and Carlo ate the rest. Which of the following statements are correct?

1 Carlo ate less than half of the chocolate bar.

2 Bella ate more than one quarter of the chocolate bar.

3 Andy and Bella ate less than three quarters of the chocolate bar altogether.

A statement 1 only

B statement 2 only

C statement 3 only

D statements 1 and 2 only

E statements 1 and 3 only

88. Andy, Bella and Carlo shared a chocolate bar.

Andy ate 40% of the bar, Bella ate 20% of the bar and Carlo ate the rest. Which of the following statements are correct?

1 Carlo ate less than half of the chocolate bar.

2 Bella ate more than one quarter of the chocolate bar.

3 Andy and Bella ate less than three quarters of the chocolate bar altogether.

A statement 1 only
B statement 2 only
C statement 3 only
D statements 1 and 2 only
E statements 1 and 3 only

89. Andy, Bella and Carlo shared a chocolate bar.

Andy ate 20% of the bar, Bella ate 25% of the remaining chocolate bar and Carlo ate the rest. Which of the following statements are correct?

1 Carlo ate more than half of the chocolate bar.

2 Bella ate more than one quarter of the chocolate bar.

3 Andy and Bella ate less than three quarters of the chocolate bar altogether.

A statement 1 only

B statement 2 only

C statement 3 only

D statements 1 and 2 only

E statements 1 and 3 only

90. Andy, Bella and Carlo shared a chocolate bar.

Andy ate 60% of the bar, Bella ate 25% of the remaining chocolate bar and Carlo ate the rest. Which of the following statements are correct?

1 Carlo ate less than half of the chocolate bar.

2 Bella ate one quarter of the chocolate bar.

3 Andy and Bella ate less than three quarters of the chocolate bar altogether.

A statement 1 only

B statement 2 only

C statement 3 only

D statements 1 and 2 only

E statements 1 and 3 only

91. In a 'magic square', each row, each column and each diagonal add up to the same total. In the magic square below, some of the numbers are missing.

4	9	2
3		
	1	▲

What is the missing number at ▲ ?

A 4

B 6

C 17

D 6

E 21

92. In a 'magic square', each row, each column and each diagonal add up to the same total. In the magic square below, some of the numbers are missing.

2	7	6
	4	■

What is the missing number at ?

A 1

B 8

C 3

D 6

E 4

93. In a 'magic square', each row, each column and each diagonal add up to the same total. In the magic square below, some of the numbers are missing.

3	6	9
3		

What is the missing number at ?

A 3
B 6
C 9
D 0
E 12

94. In a 'magic square', each row, each column and each diagonal add up to the same total. In the magic square below, some of the numbers are missing.

3%	6%	9%

3%		

What is the missing number at ?

A 3 %
B 6%
C 9%
D 0%
E 12%

95. In a 'magic square', each row, each column and each diagonal add up to the same total. In the magic square below, some of the numbers are missing.

0.030	0.060	0.090
0.030		

What is the missing number at ?

A 0.03
B 0.06
C 0.09
D 0
E 0.12

96. Chen has a spinner that is split into four equal sections.

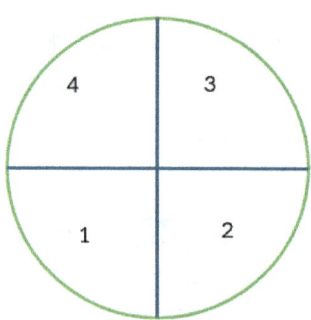

He spins the spinner, and it lands on 3. Now Emily is going to spin the spinner once. Which of these statements are correct?

X The probability of Emily getting a 2 is 1/4

Y The probability that Emily's number is less than Chen's is 1/2

Z The probability that Emily and Chen's scores add up to make more than 5 is 1/2

A none of them

B statements X and Y only

C statements X and Z only

D statements Y and Z only

E statements X, Y and Z

97.Chen has a spinner that is split into four equal sections.

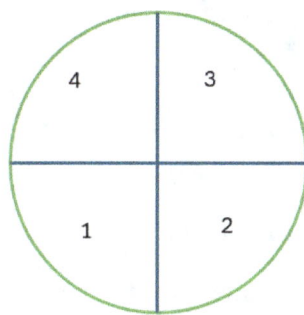

He spins the spinner and it lands on 2. Now Emily is going to spin the spinner once. Which of these statements are correct?

X The probability of Emily getting a 2 is 1/4

Y The probability that Emily's number is less than Chen's is 1/4

Z The probability that Emily and Chen's scores add up to make less than 5 is ½

A none of them

B statements X and Y only

C statements X and Z only

D statements Y and Z only

E statements X, Y and Z

98. Chen has a spinner that is split into four equal sections.

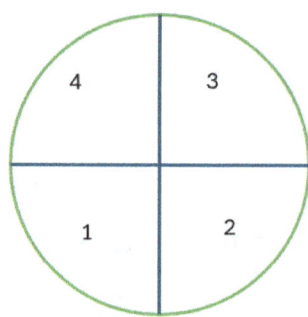

He spins the spinner and it lands on 4. Now Emily is going to spin the spinner once. Which of these statements are correct?

X The probability of Emily getting a 4 is 1/4

Y The probability that Emily's number is less than Chen's is 3/4

Z The probability that Emily and Chen's scores add up to make less than 6 is ¼

A none of them

B statements X and Y only

C statements X and Z only

D statements Y and Z only

E statements X, Y and Z

99. Chen has a spinner that is split into four equal sections.

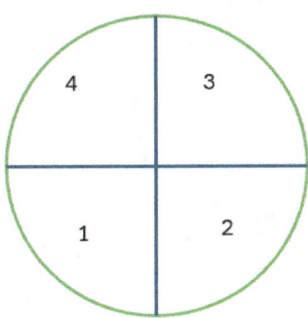

He spins the spinner, and it lands on 1. Now Emily is going to spin the spinner once. Which of these statements are correct?

X The probability of Emily getting a 1 is 1/4

Y The probability that Emily's number is less than Chen's is 0/4

Z The probability that Emily and Chen's scores add up to make more than 4 is ¼

A none of them

B statements X and Y only

C statements X and Z only

D statements Y and Z only

E statements X, Y and Z

100. Chen has a spinner that is split into four equal sections.

He spins the spinner and it lands on 4. Now Emily is going to spin the spinner once. Which of these statements are correct?

X The probability of Emily getting a 4 is 1/4

Y The probability that Emily's number is less than Chen's is 3/4

Z The probability that Emily and Chen's scores multiplied up to make more than 8 is ½

A none of them
B statements X and Y only
C statements X and Z only
D statements Y and Z only
E statements X, Y and Z

101. Ash and Beau both round the number 999 999. Ash rounds the number to the nearest million. Beau rounds the number to the nearest ten thousand. What is the difference between their answers?

A 0
B 50 000
C 100 000
D 150 000
E 200 000

102. Ash and Beau both round the number 5 849 999. Ash rounds the number to the nearest million. Beau rounds the number to the nearest ten thousand. What is the difference between their answers?

A 0
B 50 000
C 100 000
D 150 000
E 200 000

103. Ash and Beau both round the number 5 849 999. Ash rounds the number to the nearest hundred thousand. Beau rounds the number to the nearest ten thousand. What is the difference between their answers?

A 0

B 50 000

C 100 000

D 150 000

E 200 000

104. Ash and Beau both round the number 5 849 999. Ash rounds the number to the nearest thousand. Beau rounds the number to the nearest ten . What is the difference between their answers?

A 0

B 50 000

C 100 000

D 150 000

E 200 000

105. Ash and Beau both round the number 5 849 999. Ash rounds the number to the nearest million. Beau rounds the number to the nearest hundred. What is the difference between their answers?

A 0

B 50 000

C 100 000

D 150 000

E 200 000

106. Here is a graph showing the outdoor temperature during a period of 9 hours.

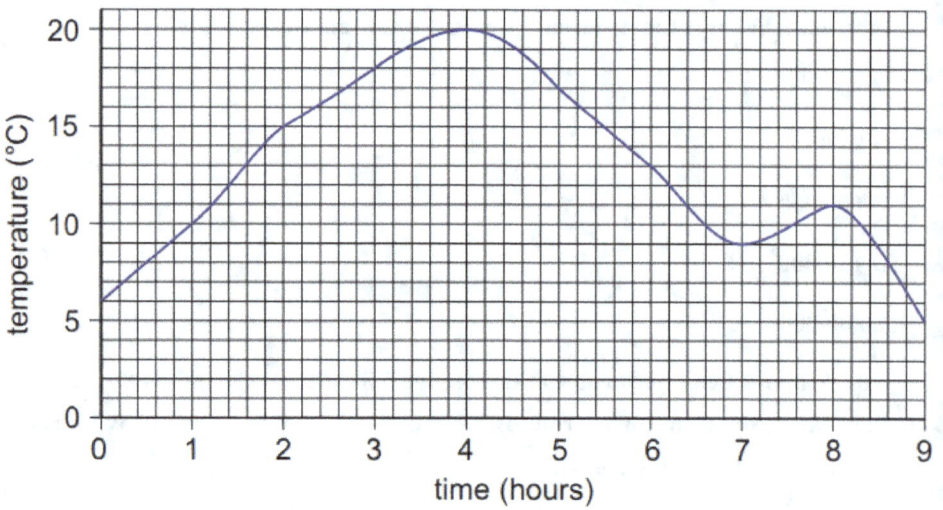

Which of these statements are true?

1 The temperature was 15°C at two different times.

2 The temperature change between 1 hour and 3 hours was greater than the temperature change between 4 hours and 6 hours.

3 The lowest temperature was 5°C.

4 Highest temperature was 20°C

A none of the statements

B statements 1 and 2 only

C statements 1 and 3 only

D statements 2 and 3 only

E statements 1, 2,3 and 4

107. Quoc bought 4 pizzas to share with Amina, Fred and Sally. Amina ate of a ¾ pizza. Fred ate 1 ¾ pizzas. Sally ate ½ of a pizza. Quoc ate all of the pizza that was left.

How much more pizza did Quoc eat than Amina?

A 1

B ¼

C ½

D ¾

E 1 ½

108. Jack bought 6 pizzas to share with John, Poppy and Sally. John ate of a ¾ pizza. Poppy ate 1 ¾ pizzas. Sally ate 1 ½ of a pizza. Jack ate all of the pizza that was left.

How much more pizza did Jack eat than John?

A 1

B 1 ¼

C ½

D ¾

E 1 ½

109. Jack bought 6 pizzas to share with John, Poppy and Sally. John ate of a ½ pizza. Poppy ate 1 ¾ pizzas. Sally ate 1 ¾ of a pizza. Jack ate all of the pizza that was left.

How much more pizza did Jack eat than John?

A 1

B ¼

C ½

D ¾

E 1 ½

110. Jack bought 7 pizzas to share with John, Poppy and Sally. John ate of a 2 ½ pizza. Poppy ate 1 ¾ pizzas. Sally ate 1 ¾ of a pizza. Jack ate all of the pizza that was left.

How much more pizza did John eat than Jack ?

A 1

B ¼

C ½

D ¾

E 1 ½

111. There are 3 boxes A, B &C with different weights.

Combined weights as below.

A+B = 10kg

A+C = 8kg

B+C = 12kg

What is the weight of A +B+C?

A 15kg

B 18kg

C 22kg

D 20kg

E 12kg

112. There are 3 boxes A, B &C with different weights.

Combined weights as below.

A+B = 20kg

A+C = 10kg

B+C = 12kg

What is the weight of A +B+C?

A 15kg

B 18kg

C 21kg

D 20kg

E 12kg

113. There are 4 boxes A, B, C &D with different weights.

Combined weights as below.

A+B+C = 20kg

A+C+D = 10kg

B+D = 12kg

What is the weight of A +B+C+D?

A 15kg

B 18kg

C 21kg

D 20kg

E 12kg

114. There are 5 boxes A, B,C,D &E with different weights.

Combined weights as below.

A+B+C+D = 20kg

A+C+D+E = 30kg

B+E = 12kg

What is the weight of A +B+C+D+E?

A 15kg

B 31kg

C 21kg

D 20kg

E 12kg

115. There are 6 boxes A, B, C,D,E &F with different weights.

Combined weights as below.

A+B+C+D+E = 20kg

A+C+D+E+F = 30kg

B+F = 12kg

What is the weight of A +B+C+D+E+F?

A 15kg

B 31kg

C 21kg

D 20kg

E 12kg

116. What is the sum of the number of faces, the number of edges and the number of vertices of a pyramid with a pentagon base?

A 18

B 20

C 24

D 22

E 38

117. What is the sum of the number of faces, the number of edges and the number of vertices of a pyramid with a rectangular base?

A 18

B 20

C 24

D 26

E 38

118. What is the sum of the number of faces, the number of edges and the number of vertices of a pyramid with a triangular base?

A 18

B 20

C 14

D 26

E 38

119. What is the sum of the number of faces, the number of edges and the number of vertices of a pyramid with a Octogen base?

A 18

B 20

C 34

D 26

E 38

120. In our number system, the value represented by a digit depends on its position. For example, in the number 273, the digit 7 represents 70. In the number 30 060, how many times larger is the value represented by the digit 3 than the value represented by the digit 6?

A 50 times

B 100 times

C 500 times

D 1000 times

E 2000 times

121. In our number system, the value represented by a digit depends on its position. For example, in the number 273, the digit 7 represents 70. In the number 300 006, how many times larger is the value represented by the digit 3 than the value represented by the digit 6?

A 50 times

B 100 times

C 50,000 times

D 500,000 times

E 1000 times

122. In our number system, the value represented by a digit depends on its position. For example, in the number 273, the digit 7 represents 70. In the number 3006, how many times larger is the value represented by the digit 3 than the value represented by the digit 6?

A 50 times
B 500 times
C 50,000 times
D 500,000 times
E 1000 times

123. In our number system, the value represented by a digit depends on its position. For example, in the number 273, the digit 7 represents 70. In the number 360,000, how many times larger is the value represented by the digit 3 than the value represented by the digit 6?

A 50 times
B 5 times
C 50,000 times
D 500,000 times
E 1000 times

124. In our number system, the value represented by a digit depends on its position. For example, in the number 273, the digit 7 represents 70. In the number 3.000,006, how many times larger is the value represented by the digit 3 than the value represented by the digit 6?

A 50 times
B 5 times
C 50,000 times
D 500,000 times
E 1000 times

125. In our number system, the value represented by a digit depends on its position. For example, in the number 273, the digit 7 represents 70. In the number 300,000,600, how many times larger is the value represented by the digit 3 than the value represented by the digit 6?

A 50 times

B 5 times

C 50,000 times

D 500,000 times

E 1000 times

126. Ash lives in Texas and she visits her cousin Beau, who lives in Sydney. Ash leaves Texas on Friday at 1300 local time on a 6-hour flight to Michigan. She spends 24 hours in Michigan. She then catches a 20-hour flight to Sydney. The time in Texas is 17 hours behind the time in Sydney. What time is it in Sydney when Ash arrives?

A Monday at 0800

B Wednesday at 1100

C Saturday at 2200

D Sunday at 1500

E Sunday at 2300

127. Ash lives in Sydney and she visits her cousin Beau, who lives in Colombo. Ash leaves Sydney on Friday at 1300 local time on a 4-hour flight to Brisbane. She spends 24 hours in Brisbane. She then catches a 10-hour flight to Colombo. The time in Colombo is 5.5 hours behind the time in Sydney. What time is it in Colombo when Ash arrives?

A Monday at 1130

B Wednesday at 1100

C Saturday at 2130

D Sunday at 2130

E Sunday at 1130

128. Ash lives in Sydney, and she visits her cousin Beau, who lives in Colombo. Ash leaves Sydney on Friday at 0100 local time on a 4-hour flight to Brisbane. She spends 12 hours in Brisbane. She then catches a 10-hour flight to Colombo. The time in Colombo is 5.5 hours behind the time in Sydney. What time is it in Colombo when Ash arrives?

A Friday at 2330
B Friday at 1130
C Saturday at 2330
D Saturday at 2130
E Friday at 2130

129. Ash lives in Sydney, and she visits her cousin Beau, who lives in Colombo. Ash leaves Sydney on Friday at 2100 local time on a 3-hour flight to KUL. She spends 6 hours in KUL. She then catches a 7-hour flight to Colombo. The time in Colombo is 5.5 hours behind the time in Sydney. What time is it in Colombo when Ash arrives?

A Friday at 0730
B Friday at 1130
C Saturday at 0730
D Saturday at 1130
E Friday at 1130

130. Ash lives in Sydney, and she visits her cousin Beau, who lives in Colombo. Ash leaves Sydney on Friday at 2100 local time on a 5-hour flight to Darwin. She spends 24 hours in Darwin. She then catches a 4-hour flight to Colombo. The time in Colombo is 5.5 hours behind the time in Sydney. What time is it in Colombo when Ash arrives?

A Friday at 1230
B Friday at 0030
C Saturday at 0030
D Saturday at 1230
E Sunday at 0030

131. Here is a shape made of black and white parts:

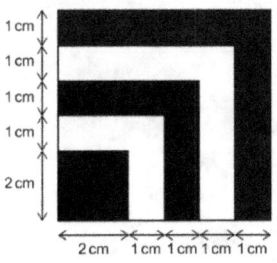

[diagram not to scale]

What is the total area that is black?

A 14 cm2

B 15 cm2

C 20 cm2

D 22 cm2

E 24 cm2

132. Here is a shape made of black and white parts:

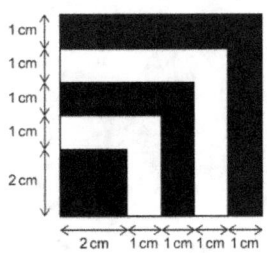

[diagram not to scale]

What is the total area that is white?

A 14 cm2

B 15 cm2

C 20 cm2

D 22 cm2

E 24 cm2

133. Jan is making a mixed fruit drink for a party. She puts some orange juice into the jug, as shown.

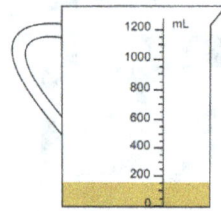

She then adds 6 times as much apple juice. How many millilitres of fruit drink does Jan make?

A 400 mL

B 525 mL

C 900 mL

D 1050 mL

E 1500 mL

134.

Through what angle is it rotated anticlockwise?

A 60°

B 120°

C 180°

D 240°

E 300°

135. Babybah's little brother RIZZY tried to draw three different nets of a cube, but he got it wrong. He only drew four squares for each net, but a complete cube net has six squares. Here are his three drawings

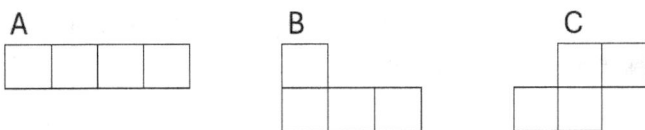

Babybah wants to help, by adding two more squares to each net to make a correct cube net. However, he is not sure if this is possible for all of the drawings. Which of the three drawings can be made into complete cube nets by adding two squares?

A drawing 1 only

B drawing 2 only

C drawing 3 only

D drawings 1 and 3 only

E drawings 1, 2 and 3

136. Babybah's little brother RIZZY tried to draw three different nets of a cube, but he got it wrong. He drew seven squares for each net, but a complete cube net has six squares. Here are his three drawings

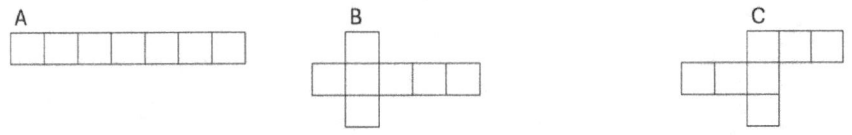

Babybah wants to help, by removing 1 square to each net to make a correct cube net. However, he is not sure if this is possible for all of the drawings. Which of the three drawings can be made into complete cube nets by removing one square?

A drawing 1 only

B drawing 2 only

C drawing 3 only

D drawings 2 and 3 only

E drawings 1, 2 and 3

137. Babybah's little brother RIZZY tried to draw three different nets of a cube, but he got it wrong. He drew eight squares for each net, but a complete cube net has six squares. Here are his three drawings

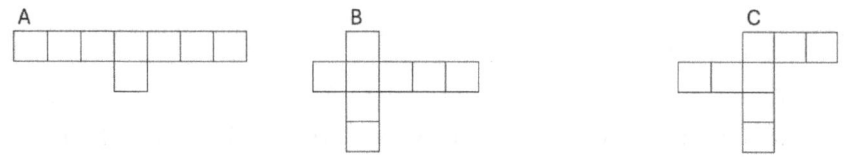

Babybah wants to help, by removing 2 squares to each net to make a correct cube net. However, he is not sure if this is possible for all of the drawings. Which of the three drawings can be made into complete cube nets by adding one square?

A drawing 1 only

B drawing 2 only

C drawing 3 only

D drawings 2 and 3 only

E drawings 1, 2 and 3

138.

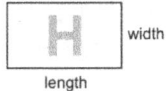
width
length

Harry is making shapes out of identical rectangular cards with the letter H on. He makes shape X, then adds another card to make shape Y, as shown.

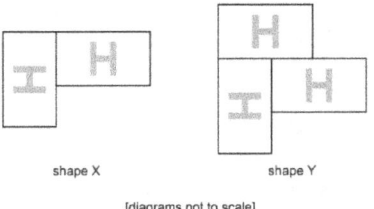

shape X   shape Y

[diagrams not to scale]

The perimeter of shape X is 64 cm and the perimeter of shape Y is 82 cm. What is the width of each of Harry's rectangles?

A 3 cm

B 4 cm

C 5 cm

D 8 cm

E 9 cm

139.

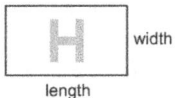
width
length

Harry is making shapes out of identical rectangular cards with the letter H on. He makes shape X, then adds another card to make shape Y, as shown.

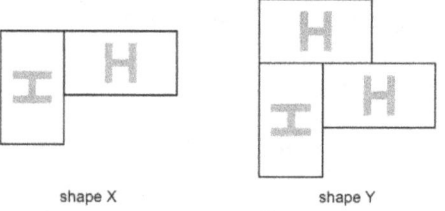

shape X   shape Y

[diagrams not to scale]

The perimeter of shape X is 42cm and the perimeter of shape Y is 52 cm. What is the width of each of Harry's rectangles?

A 3 cm

B 4 cm

C 5 cm

D 8 cm

E 9 cm

140. Claudia makes a number pattern using these rules: • Choose the first three numbers. • After this, each number is the sum of the three numbers before it. Here is Claudia's pattern with three missing numbers labelled X, Y and Z.

2, X, 0, Y, 6, Z All the numbers are whole numbers. What number does Z represent?

A 8

B 9

C 10

D 11

E 12

141. Claudia makes a number pattern using these rules: • Choose the first three numbers. • After this, each number is the sum of the three numbers before it. Here is Claudia's pattern with three missing numbers labelled X, Y and Z.

4, X, 2, Y, 12, Z All the numbers are whole numbers. What number does Z represent?

A 8

B 9

C 10

D 11

E 22

142.Claudia makes a number pattern using these rules: • Choose the first two numbers. • After this, each number is the sum of the two numbers before it. Here is Claudia's pattern with three missing numbers labelled X, Y and Z.

4, X, Y, 12, Z All the numbers are whole numbers. What number does Z represent?

A 8
B 9
C 20
D 11
E 22

143.Claudia makes a number pattern using these rules: • Choose the first two numbers. • After this, each number is the subtraction of the two numbers before it. Here is Claudia's pattern with three missing numbers labelled X, Y and Z.

20, X, Y, 4, Z All the numbers are whole numbers. What number does Z represent?

A 8
B 9
C 4
D 11
E 22

144.In a game, I collect stars and hearts. I get points when I collect a star. The number of points for a star is always the same. The number of points for a heart is always the same. If I collect 3 stars and 5 hearts, I get 27 points. = 27 points If I collect 5 stars and 7 hearts, I get 41 points. = 41 points How many points do I get if I collect 1 star and 1 heart?

A 7

B 12

C 14

D 24

E 50

145. How many whole numbers between 1 and 100 are multiples of 3, but are not multiples of 8? You should include 3 itself.

A 21

B 25

C 27

D 29

E 33

146. How many whole numbers between 1 and 100 are multiples of 4, but are not multiples of 8? You should include 4 itself.

A 10

B 25

C 12

D 29

E 33

147. How many whole numbers between 1 and 100 are multiples of 5, but are not multiples of 9? You should include 5 itself.

A 18

B 16

C 14

D 29

E 33

148. How many whole numbers between 1 and 100 are multiples of 7, but are not multiples of 9? You should include 7 itself.

A 21

B 25

C 13

D 29

E 33

149. How many whole numbers between 1 and 100 are multiples of 2, but are not multiples of 9? You should include 2 itself.

A 21

B 25

C 45

D 29

E 33

150. How many whole numbers between 1 and 100 are multiples of 4, but are not multiples of 6? You should include 4 itself.

A 21

B 25

C 17

D 29

E 33

151. Katy has blue, green, red and yellow balls in a bag. There are ¼ as many red balls as blue balls. If she takes one ball out of the bag without looking:

• the probability that it is blue is 0.4

• the probability that it is green is 0.3

The shed added number of yellow balls so the number of yellow balls equals the blue balls.

What is the number of balls in the bag once additional red balls were added?

A 0.1

B 0.2

C 0.3

D 0.4

E 0.5

152. Here is a graph showing the outdoor temperature during a period of 9 hours.

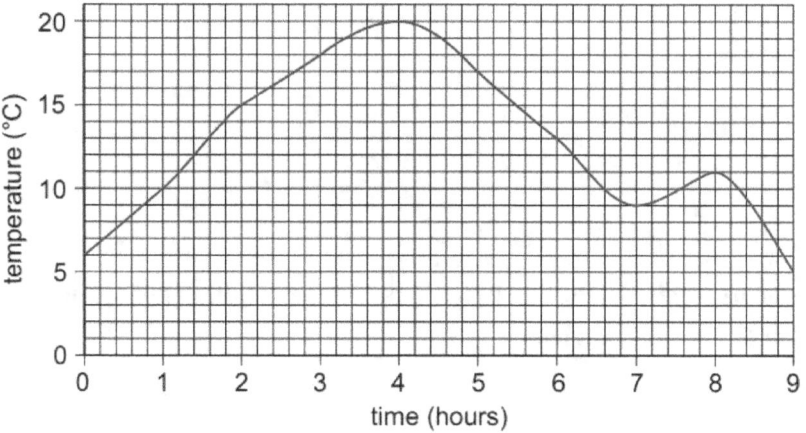

Which of these statements are true?

1 The temperature was 10°C at four different times.

2 The temperature change between 0 hour and 4 hours was lower than the temperature change between 4 hours and 9 hours.

3 The difference between the lowest temperature and the highest temperature was 15°C.

4 The difference in temperature between 0 hour and 9 hour was 1°C.

A none of the statements
B statements 1 and 2 only
C statements 1 and 3 only
D statements 2 and 3 only
E statements 1, 2,3 and 4

153. There is a rectangle with sides 9cm and 15cm. How many squares can be made to have the same area in total if the side of each square is 3cm.

A 2

B 15

C 40

D 4

E 5

BLANK

BLANK

BLANK

Answers

1	E
2	E
3	B
4	D
5	C
6	D
7	A
8	B
9	A
10	A
11	B
12	A
13	C
14	A
15	C
16	A
17	A
18	A
19	A
20	A
21	A
22	D
23	B
24	A
25	D
26	B

27	C
28	A
29	A
30	A
31	A
32	B
33	B
34	D
35	D
36	A
37	D
38	A
39	A
40	B
41	C
42	C
43	A
44	A
45	B
46	A
47	E
48	A
49	A
50	C
51	A
52	A
53	A
54	C
55	D
56	E
57	A
58	E
59	B
60	A
61	E
62	E
63	E
64	E
65	E
66	B
67	D
68	C

69	E
70	D
71	C
72	D
73	D
74	A
75	C
76	E
77	A
78	A
79	A
80	A
81	E
82	A
83	B
84	A
85	B
86	B
87	E
88	E
89	E
90	D
91	D
92	B
93	B
94	B
95	B
96	E
97	E
98	E
99	E
100	E
101	A
102	D
103	B
104	A
105	D
106	E
107	A
108	B
109	E
110	E

111	A
112	C
113	C
114	**B**
115	B
116	D
117	A
118	C
119	C
120	A
121	C
122	B
123	B
124	D
125	D
126	A
127	C
128	E
129	C
130	E
131	D
132	B
133	D
134	B
135	E
136	D
137	D
138	E
139	C
140	B
141	E
142	C
143	C
144	A
145	D
146	C
147	A
148	C
149	C
150	C
151	B
152	E

Burnout Breakthrough

Effective Strategies to Manage Stress and Thrive in a Demanding World

Harmony Royce

Copyright © 2024 Harmony Royce

All rights reserved.

www.ingramcontent.com/pod-product-compliance
Lightning Source LLC
Chambersburg PA
CBHW071109240526
45469CB00006BD/2403